Beyond Curiosity

Dahlia Breeding Through the Seasons

Written, Photographed & Designed by
Rebecca McConnell

Spring

Summer

Autumn

Winter

Farm News

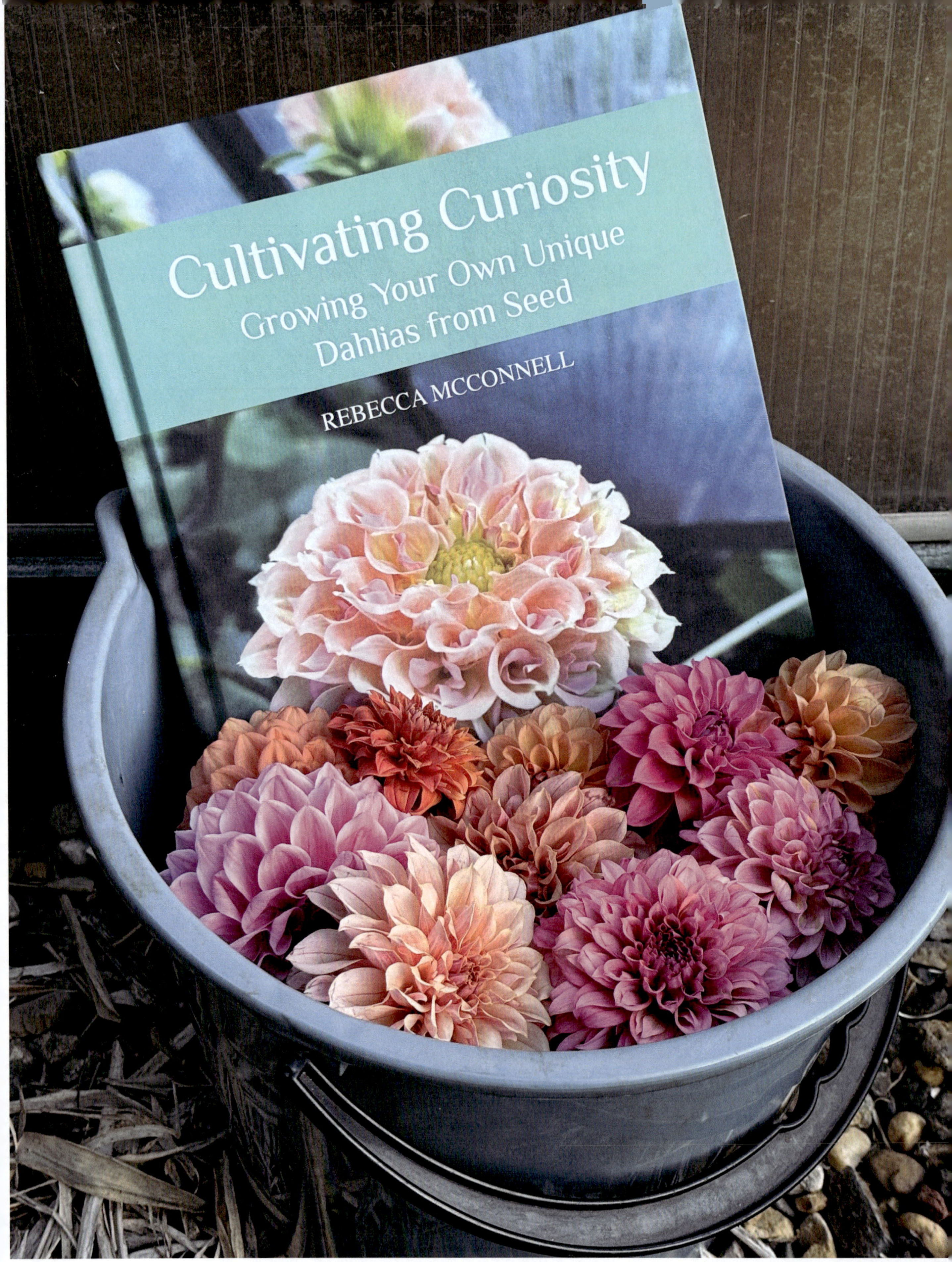
Cultivating Curiosity
Growing Your Own Unique
Dahlias from Seed
REBECCA MCCONNELL

When I was penning "Cultivating Curiosity", and sorting through tens of thousands of seedling photos, designing the layout, navigating print and publishing options, and feeling the trepidations of putting myself out into the world, I never imagined the wonderful opportunities that my self-published dahlia breeding book would create.

I am grateful to everyone who purchased a copy, and the affirming feedback I've received. It is such a buzz knowing that there are readers all over the world, from Sweden, the UK, Canada & US, Hungary, Ireland, Germany, New Zealand and many of my Australian dahlia community, flipping through those carefully crafted pages.

Even more so, it warms my heart hearing that my book has given growers the confidence to collect their own dahlia seed, nurture plants to maturity and delight in their own unique creations.

Since publishing, I've noticed a surge of interest not only in dahlia hybridisation generally, but specifically in our Serenade Farm dahlia breeding program. Our following on Instagram, and in-person interactions, affirm to me that there is interest in the day-to-day discoveries, decisions and strategies that we implement here. I enjoyed posting insights about how the principles from "Cultivating Curiosity" apply on my own farm throughout the season. As I reflected back, I realised returning to the print format would more cohesively document the discoveries that unfolded in the Serenade breeding program 2026. Alongside the practical, I've also woven in more science principles underpinning the traits that I've chosen to hybridise for, and the specific relevance to my breeding lines.

I hope that this next instalment of the Serenade Dahlias journey further piques your interest in this wondrous process of hybridisation and empowers you to cultivate your curiosity into full bloom.

Love Bec x

A dahlia breeding journey takes time and patience. It is a continuous process that shifts focus through the seasons. This rise and fall of growth and dormancy is reflected in our hybridisation practices.

We'll begin in Spring, with planning, planting and propagation. Spring is ripe with potential - when we invest our time and energy for what is to come. We'll discuss the layout of Serenade Farm and our preparations for the coming season. Through this section of the book we examine the impact of daylength, as well as what other physiology and key conditions trigger tubers to break dormancy and those that cue dahlia seeds to germinate.

Summer is where it gets fun - growth, blooms and exploring the breeding lines. For the visual learners, I've packed in plenty of photos of this year's new patch of 2000 first-year seedlings, alongside images of their seed parents and siblings. For those of us who love to geek out, I'm also exploring pigment pathways for petal colour, and also how I've begun to classify my petaloid lines to refine this trait with more nuance.

Autumn is when dahlias are at their peak. We had a tremendous Autumn this year with our cut flower business in full swing, harvesting every single morning until early April, and I spent almost every afternoon assessing, photographing and documenting the first to fourth year seedlings, and deciding which will be kept for the next year. We had several weeks of relatively fine weather through April. This really was a boon to my hybridisation process, allowing the most successful hand pollination to date (since April, which is mid Autumn, is normally our wettest time of year). And then May is when we harvest, dry and store our beautiful dahlia seed, holding all the potential for next year's dahlias.

And Winter? Winter is the time we consolidate. We connect with others. We reflect on what was discovered and what was learnt. On how our goals may have shifted and changed. In Winter, we dream.

Spring

Planning and layout

As happens on a lot of properties, gardens evolve. Our Serenade Farm dahlia beds were re-purposed from other growing we had done in the past. Shifting out of a vegetable market garden set-up, we had already built up our soil health and established infrastructure including trellises (originally used for beans & cucumbers but now an easy support system for tall growing dahlias) and micro-sprayer irrigation.

As such, we have three distinct dahlia growing areas. There is not enough space between them to completely mitigate cross pollination - certainly the bees and pollinators can easily travel from one area to the other. However, when observing the bees at work, it's obvious that they are more likely to visit blooms that are in a row and clumped together. So the natural separation can be utilised along with other intentional planning to increase our odds for the crosses we would prefer. It's crucial though to plan this in Spring when planting.

All of our growing areas are planted fairly intensively, (80-90cm wide beds with 60cm paths). The beds are slightly raised, essential here in wet years since dahlias hate to have their roots in sodden soil. When deciding which dahlias to plant where, we separate into the three areas based on the primary purpose.

Our Main Cutting Patch in the centre is for our cut flower production needs. This patch needs easy access for harvesting every morning, and we grow large numbers of each cultivar to provide consistency to our florists. It includes some reliable, named cultivars, but most of this patch is third, fourth and fifth year Serenade seedlings being assessed for cut flower production. By growing so many side by side with tested cultivars, it gives opportunity to do a comparative assessment. It is obvious if productivity is relatively poor, and reveals any genetic health issues, (such as susceptibility to fungal issues or particular pests) if plants in adjacent rows are still healthy. The only plant supports in this area are corralling ropes, so the plants also need to be self-supporting.

The other advantage of planting all the best cut flower cultivars together (and loosely in colour blocks) is that this naturally optimises the best open pollination for the cut flower seed lines. Any seedlings in this area that have not performed simply get cut back before leaving flowers on the plants for pollination.

The proving ground to the West is primarily the second year seedling assessment area. It is protected from harsh western sun by neighbour's trees, and has the original trellis set up. Tubers are planted either side of vertical mesh (10x10cm / 4x4" wire), with twine looped around plants to keep them upright. They can slip through the assessment criteria of plant habit (self-supporting) assessment, because they have more support.

The soil holds more moisture in this area, and also has irrigation installed, which also means these plants don't get stress tested in hot dry seasons.

As opposed to the very practical assessment criteria that I'm using in the main cutting patch, I conceptualise this area as more of a living, evolving brainstorm. These seedlings are the less conventional forms, unique seedlings that I am keeping more for breeding curiosity than their specific potential for cut flowers.

This season, sections of this patch also doubled as hybridisation hubs, with several adjacent rows planted strategically with my petaloid lines, and other unusual forms. I interplanted with my "Dancing in the Moonlight" lines that bring in the functional strengths and also some of the "Whisper" seedlings.

Bucket of Serenade seedlings

The best "real estate" on the farm sits in the two patches on the Eastern side. This is where we've been working on the soil the longest and drainage is best, and for this reason, it is where I planted the 2000 new first-year seedlings. The good drainage will also ensure that in a particularly wet season they are less at risk of being lost to rot, which is crucial when it is the only plant that exists with that specific DNA anywhere.

The first year seedlings are transplanted with very tight spacing of 10-15cm (4-6") because I intend to remove many of them through the season after assessing the first few flushes of blooms. Although I am wanting them to grow as well as possible (best soil, best sun) I am still trying to pack in as many as I can so I have ample to choose from. This allows me to be fussy, accepting the risk that some removed plants might have performed better given more space.

We do make tweaks to the planting layout each year. This season, I allocated two beds to my favourite second year cut flower seedlings in the Eastern patches. I gave them second year assessment while keeping some hybridisation distance from the artsy, creative weirdos in the Western beds.

Soil preparation and health

In order to assess our dahlia seedlings accurately it is essential that we grow them well to begin with. We give them a full sun location and time our planting to align with seasonal and daylength cues. We provide optimal temperatures and moisture levels where we can to avoid undue stress. Nutrition is the other key aspect to plant health and so during Spring we assess our soil and prepare it for the coming season.

Last Winter we had a laboratory soil test done so that we could be more precise with our amendments and fertiliser schedule. Finding a lab willing to deal with such a small-scale customer proved more challenging than expected. This was further complicated by our location in a fire ant biosecurity zone. However, it was well worth the effort and cost, as we learnt that our soil not only had ample nutrition and extremely high levels of organic matter, but also great bio-availability of those nutrients. The results indicated there was no need to add fertiliser at all. It validated all the hard work over the years focussing on soil health, regenerative principles and microbial diversity.

Previously, on planting, we would add a balanced organic food (in pellet form) which would slowly release through early stages of growth, and then would add both organic potassium and organic phosphorus products later (from budding stage onwards). Instead, the idea was that this year we would allow the plants to utilise the nutrition from the soil, and continue to observe them for any signs of deficiency (which fortunately did not arise).

However, there were a few areas within the main cutting patch that still needed a bit of love after several seasons of excessive rain that had compacted the soil structure.

Over winter I planted daikon radish in some of the worst beds to help break up and aerate the soil. Since the dahlia tubers in that bed had rotted there was no need to dig them up, and the daikon seeds were direct sown. It was even more effective than I had expected, and as a bonus we had some daikon radish to eat and ferment, and the tops were simply left as mulch over the soil to break down. Other smaller blocks where I had decided to pull particular cultivars (either possible virus, or in some cases cultivars that simply didn't perform well enough to keep), I popped in a quick winter cover crop of mustard, which I only allowed to grow to just before flowering stage, then chopped and dropped, leaving the greens to break down on top of those beds.

In early Spring I also addressed an area in this patch where I had discovered some mild evidence of root knot nematodes. I cleared out (and disposed of) the remaining tubers, just in case. I then soaked the soil with a bio-active compost tea to add in extra diversity, and direct sowed a mix of mustard and African Marigold seeds which can help manage mild infestations. I still don't intend to plant any dahlias (or any type of potential host plants) in that bed this season, and may just plant marigolds again for another season just to be safe.

Besides these special needs areas, the rest of the farm housed our tuber clumps over winter and early spring, happily overwintered under the soil. Weeding was the main consideration for those areas until the digging began.

Dig/divide/replant process

Since we overwintered all dahlias in the ground, we avoided the bother of storing. In Spring, everything was systematically lifted and processed. Depending on whether I decided to keep and regrow the cultivar/seedling, or let it go, the tuber clump was either split into smaller chunks and then replanted into its new position or sold off via our roadside honesty cart. Whilst I dug, my husband would divide clumps as well as adding the new thick layer of mulch over the freshly dug beds, and I would then replant, usually within a few days, unless I was moving cultivars around. This whole process, done gradually over the farm, takes several months.

Spring mulching

Each year, we cover both the beds and paths in a very thick layer (100mm) of wood chip mulch. Multi-functional, this suppresses weed growth year round, reduces moisture loss in summer, and feeds the soil life and adds a tremendous amount of organic mass to the soil as it breaks down each season. We haven't had any issues with nitrogen draw-down, slugs/snails or widespread fungal issues. Those are the three common contraindications to using wood chip around the base of plants, reported by other growers who don't like this approach, which again, reinforces the idea of finding what works for you in your own specific conditions and situation.

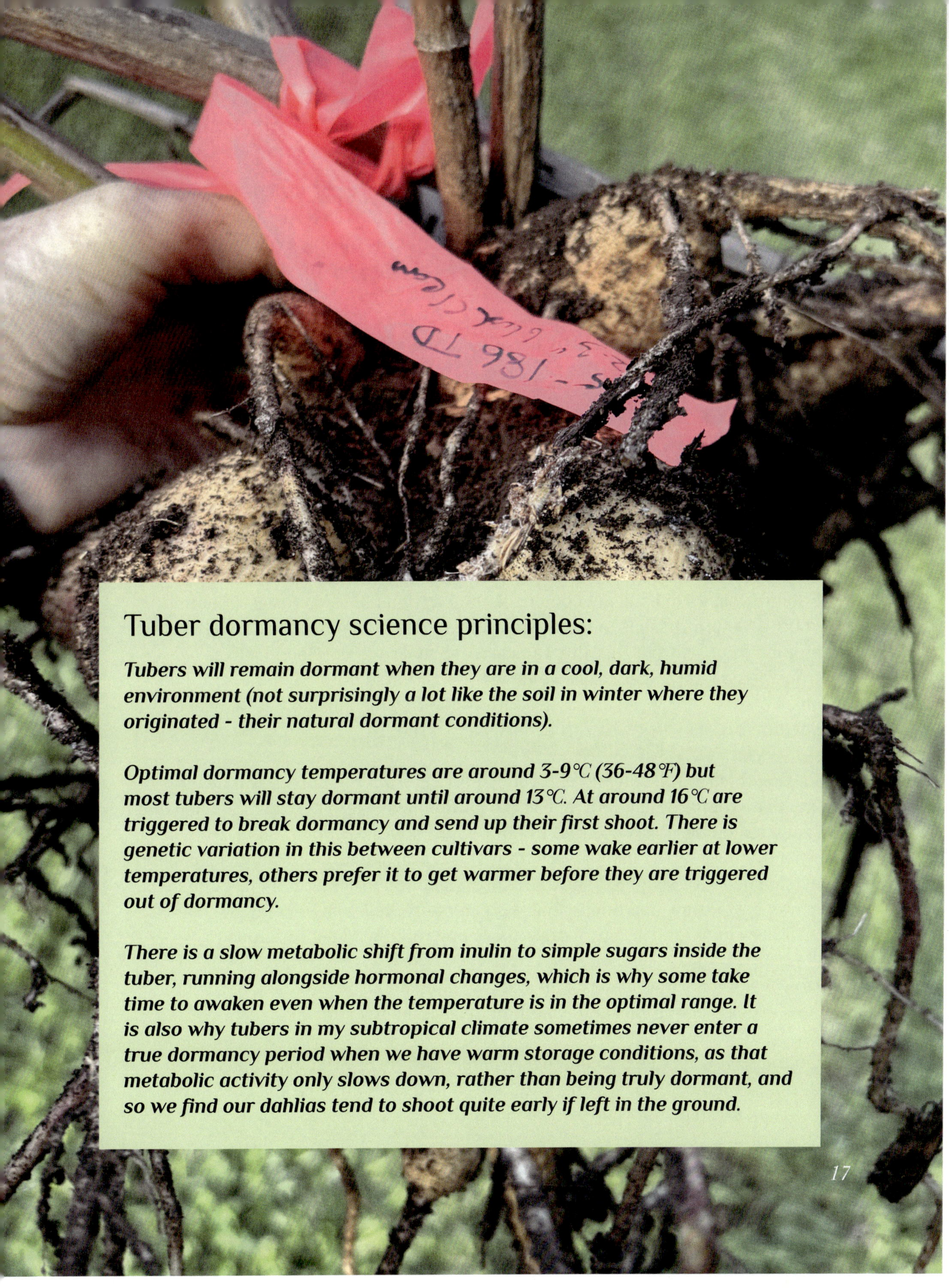

Tuber dormancy science principles:

Tubers will remain dormant when they are in a cool, dark, humid environment (not surprisingly a lot like the soil in winter where they originated - their natural dormant conditions).

Optimal dormancy temperatures are around 3-9℃ (36-48℉) but most tubers will stay dormant until around 13℃. At around 16℃ are triggered to break dormancy and send up their first shoot. There is genetic variation in this between cultivars - some wake earlier at lower temperatures, others prefer it to get warmer before they are triggered out of dormancy.

There is a slow metabolic shift from inulin to simple sugars inside the tuber, running alongside hormonal changes, which is why some take time to awaken even when the temperature is in the optimal range. It is also why tubers in my subtropical climate sometimes never enter a true dormancy period when we have warm storage conditions, as that metabolic activity only slows down, rather than being truly dormant, and so we find our dahlias tend to shoot quite early if left in the ground.

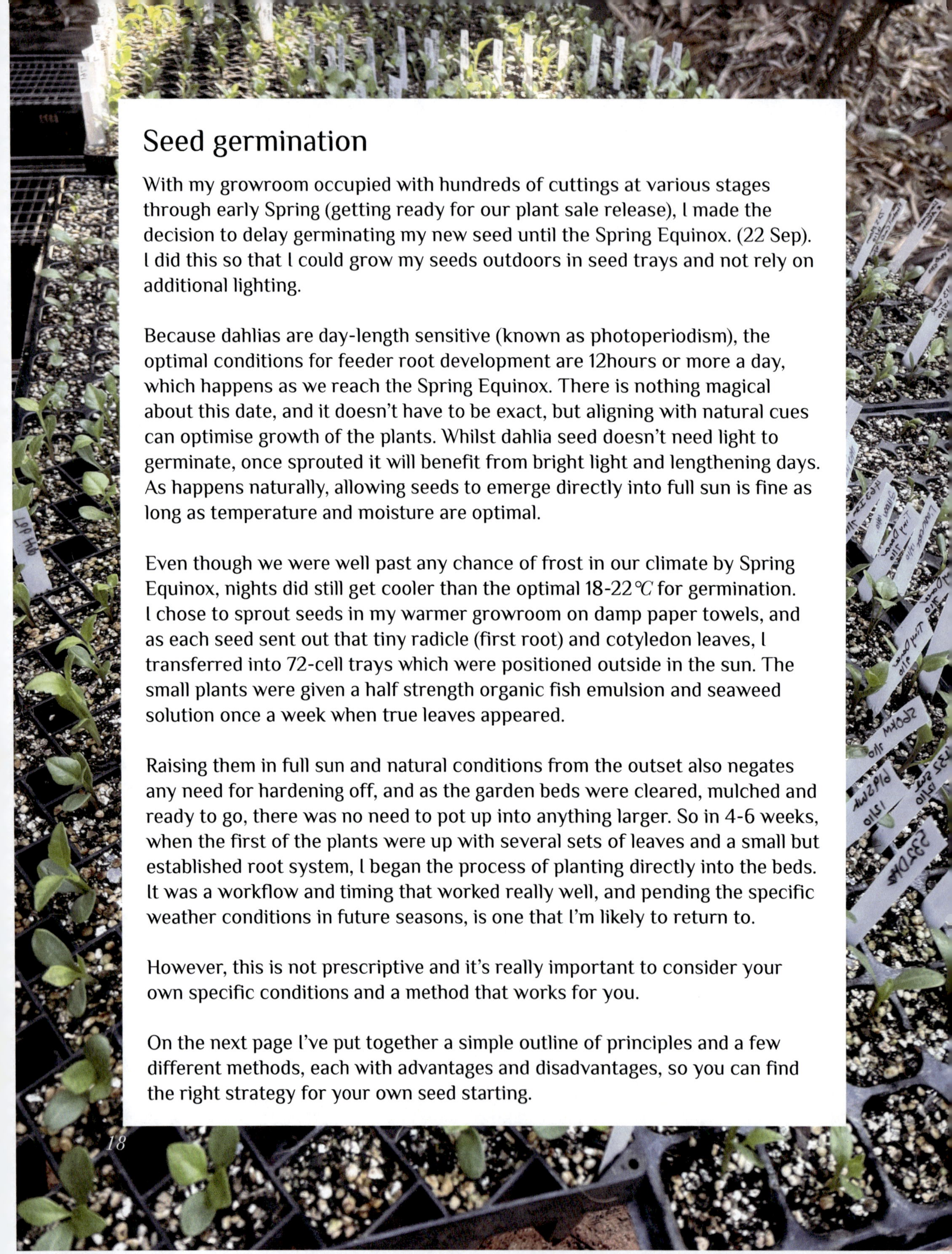

Seed germination

With my growroom occupied with hundreds of cuttings at various stages through early Spring (getting ready for our plant sale release), I made the decision to delay germinating my new seed until the Spring Equinox. (22 Sep). I did this so that I could grow my seeds outdoors in seed trays and not rely on additional lighting.

Because dahlias are day-length sensitive (known as photoperiodism), the optimal conditions for feeder root development are 12hours or more a day, which happens as we reach the Spring Equinox. There is nothing magical about this date, and it doesn't have to be exact, but aligning with natural cues can optimise growth of the plants. Whilst dahlia seed doesn't need light to germinate, once sprouted it will benefit from bright light and lengthening days. As happens naturally, allowing seeds to emerge directly into full sun is fine as long as temperature and moisture are optimal.

Even though we were well past any chance of frost in our climate by Spring Equinox, nights did still get cooler than the optimal 18-22℃ for germination. I chose to sprout seeds in my warmer growroom on damp paper towels, and as each seed sent out that tiny radicle (first root) and cotyledon leaves, I transferred into 72-cell trays which were positioned outside in the sun. The small plants were given a half strength organic fish emulsion and seaweed solution once a week when true leaves appeared.

Raising them in full sun and natural conditions from the outset also negates any need for hardening off, and as the garden beds were cleared, mulched and ready to go, there was no need to pot up into anything larger. So in 4-6 weeks, when the first of the plants were up with several sets of leaves and a small but established root system, I began the process of planting directly into the beds. It was a workflow and timing that worked really well, and pending the specific weather conditions in future seasons, is one that I'm likely to return to.

However, this is not prescriptive and it's really important to consider your own specific conditions and a method that works for you.

On the next page I've put together a simple outline of principles and a few different methods, each with advantages and disadvantages, so you can find the right strategy for your own seed starting.

Germinating Dahlia Seeds

Temperature
Ideal range 18-22℃ (65-72℉)

Moisture & Air
Consistently damp NOT soggy medium
Seeds need oxygen to avoid rot & damping off

Light & Dark
Light not needed to germinate, but strong light once sprouted, full sun fine.
Plant 5-10mm deep if direct sowing

Daylength
12hr+ days encourage feeder root growth
Supplement lighting if germinating before Spring Equinox

No nutrition needed until true leaves emerge. If starting in sterile mix, weak foliar feed each week until transplanted

A few different options:

Method	*Best for*	*Advantages*	*Disadvantages*
Direct sow in garden beds	Own seed and optimal climates (after frost)	Fast and easy	Slugs, heavy rain, uneven germination
Scatter into trays/flats	Growing many seedlings	Efficient use of space and time	Requires pricking out
Individual cells/pots	Beginners wanting simplicity	Minimal root disturbance	Uses more space & seed raising mix
Paper towel pre-sprouting	Valuable seed & controlling timing	Lets you monitor viability	Delicate handling & time consuming

DIRECT SOW
1. Prepare loose, weed-free soil
2. Sow 5–10mm deep
3. Keep evenly moist until emergence

INDIVIDUAL CELLS OR POTS
1. Sow 1 seed per cell of moist mix
2. Cover with mix or vermiculite

SCATTER INTO FLATS
1. Scatter seed across moist seed mix
2. Cover with mix or vermiculite
3. Prick out after first true leaves appear

PAPER TOWEL PRE-SPROUTING
1. Place seeds in moist paper towel 18-22℃
2. Check daily for sprouting
3. Transfer to cells when roots emerge

Summer

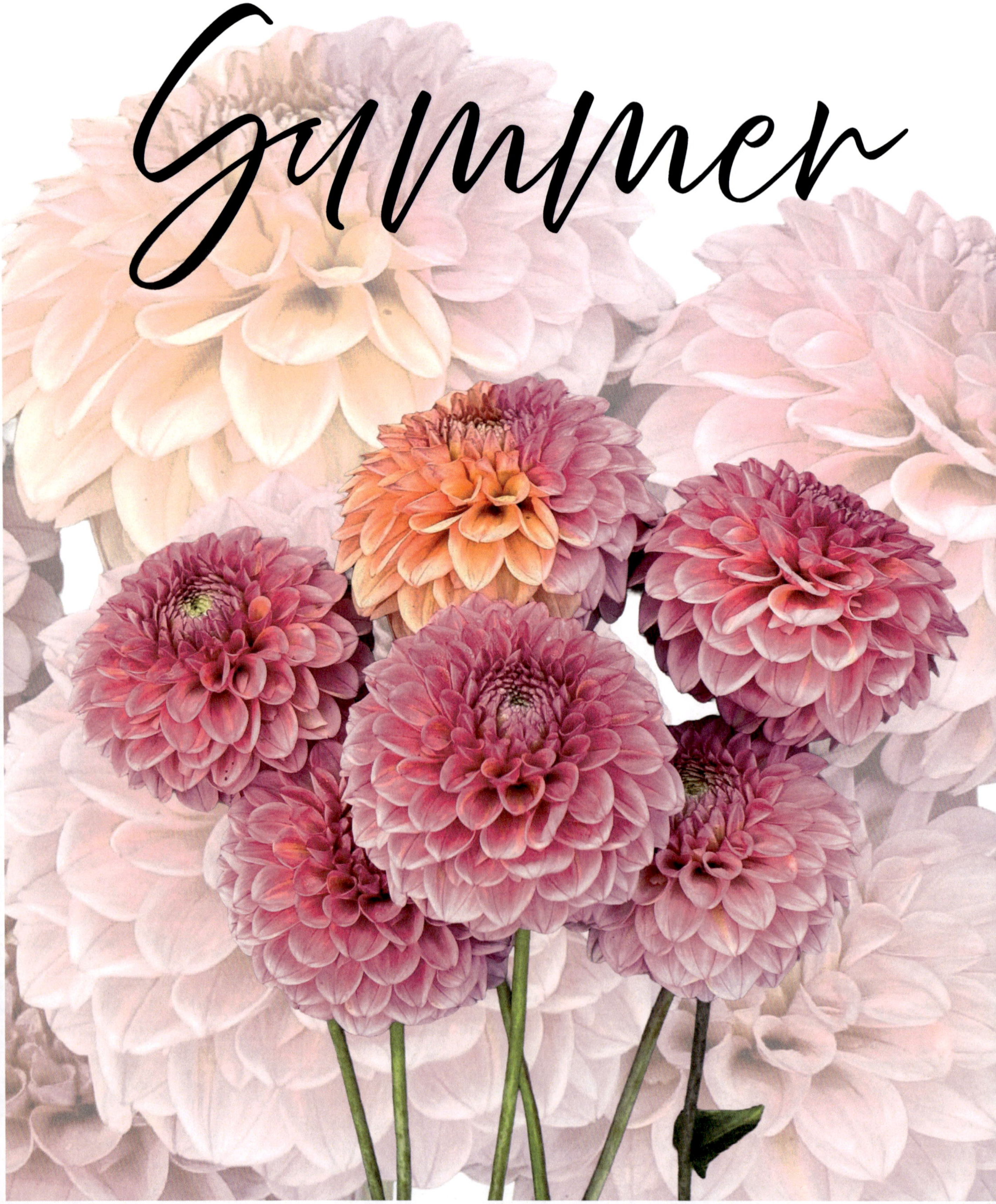

Growing in a harsh climate

Summer can be harsh and unpredictable in our subtropical climate here on Tamborine Mountain, so I breed for resilience in my dahlias.

This season was better than the previous five. With big swings in temperature and rainfall, we were around 3-4 weeks later than usual with production, but that was a minor issue.

The previous year we had a tropical cyclone in March, which was the first in our area of that severity in 50 years. The year before that, a freak tornado ripped through on Christmas night, despite that not being a weather system ever heard of in our area. The two years before it was record breaking rains and flooding. The year before that? We had such dry weather that it resulted in bushfires in the region. Flower farming is not for the faint of heart.

This highlights the need to assess over many seasons, and rather than suggesting that we need to go easy on every seedling and excuse the faults because of atypical weather, perhaps we should be breeding dahlias that have a high degree of resilience and perform well in all sorts of conditions? Most especially, it reinforces one of my underlying breeding goals, which is to develop strong healthy dahlias that are genetically adapted to our specific conditions.

Breeding goals coming to life

There are two quite different arms to my Serenade dahlia breeding program. On one hand, I am breeding exceptional, productive cut flowers, that are climate adapted to use for our florist orders. Functional traits are the most important in these lines, and colours need to reflect those requirements which tend to shift alongside styling trends, both in retail floristry and their wedding and event work.

In parallel, I am also exploring new forms, pushing boundaries, and learning as much as possible about what dahlia genetics are capable of creating. These lines are curiosity driven, experimental, and I think of them more like art than engineering. With these seedlings I will sometimes keep something that does not rate high on every functional criteria. However, these seedlings are only one move in a long game of building new aesthetics into functionally excellent blooms. I would not release an odd or quirky seedling until I was also happy with it functionally as well, and I do expect that to take many years to achieve.

Aesthetic vs functional traits

In order to not muddy my assessment between the two contrasting objectives, I separate out aesthetic considerations, which are largely subjective, and the more objective functional ones.

The mistake new dahlia breeders commonly make is keeping every pretty flower. It becomes easier over time to let go, as we become more discerning. I'm now comfortable growing the majority of my first year seedlings as annuals. They are enjoyed for one season, and then passed on to someone else or added to the compost.

This year my breeding criteria tightened up more than ever in regard to the functional criteria. I would not graduate a seedling within my cut flower lines unless it had all four of the "S-A-P-C" (Stem, Attachment, Position, Centre) traits - an exceptional Stem, secure bloom attachment, presenting at a 45degree angle or higher, and high petal count with a good centre. The first 3 are non-negotiable and I can typically assess after 2-3 flushes of blooms. The final one I do give some grace to hard centres through the hottest humid months, but I do need to see them improve significantly by March.

Other functional aspects that culled for this season included

- Tendency for whites to discolour or soft petals that bruise easily
- Less vigorous or less healthy plants
- Bright orangey reds, fluorescent yellows or highly saturated cool pink (unless they had something else truly exceptional about them)
- Loose open decorative forms with relatively poor vase life

Specific to my second-fifth year seedlings that were in the main cutting patch, I was also looking for productivity, reliability, resilience

- Consistent and frequent flushes of blooms to give good regular harvests
- High number of harvestable stems
- Low maintenance plants that didn't have a diva attitude, resilient to fungal and pest issues in our climate.

So let's take a closer look at the lines that I'm currently developing in the Serenade dahlia breeding program. Some are still focussed on the aesthetics, others are ticking all those cut flower assessment goals.

For curiosity's sake - the petaloid developments

Double dahlias with petaloids is the Serenade "signature". Of my breeding lines, this generates the most interest, so let's begin by unpacking how this is evolving.

I'm increasingly confident in my initial hunch that petaloids are not uncommon in dahlia genetics. In the past, they have been selected against by show growers who view them as a fault. The only place for them in exhibition standards has been as a single row collar in the collerette form.

The more I select for this trait within my fully doubles, the more it comes through, with well over 50% of blooms presenting with petaloids in some seed parent blocks in the first year patch this year. This frequency means I can now be extremely fussy about which to keep for breeding.

What really IS a petaloid anyway?

Petaloids are the small petal-like structure appearing inside the main petals (ray florets). They have evolved from the reproductive parts of the floret called stamens. These so called “staminodes” (non-reproductive stamen) differentiate into petal tissue and become laminar (flat and blade-like) and pigmented so that they look more like mini-petals than the skinny, yellow stamen structure they evolved from.

It makes sense that biologically, variations arise from the degree to which the stamens have been modified into petaloids. If they are only partially converted, they are smaller and thinner, and less petal-like. Contrasting colour between the petals and petaloids also speaks to this incomplete conversion, where the stamens didn’t originally have coding for the pigment pathways to develop certain colours that the petal behind them do.

I'm separating my petaloid seedlings into different classes based on nuance of the petaloids and their relationship to the petal that envelopes them.

SIZE: Petaloid size relative to the petal behind them. Are they tiny structures, barely visible in the centre of the petal? Or a large flourish, in some cases extending beyond the edge of the main petal?

COLOURING: The majority have the same pigmentation in both petal and petaloid, but what I'm really hoping to develop more is the contrasting colour.

SHAPE & CURVATURE: In some cases the main petal cups up in an involute curve and cradles the inner petaloid; in others, both petal and petaloid wrap back flat to give full visibility to both.

In my Serenade petaloid lines, these following classes are still blurred and overlapping with variations not particularly defined, since I'm still in an early stage of the process. This winter I'll decide on my priorities, how to plant for intentional open pollination and also which to hand cross and hope to see more separation.

1. Contrasting colour petaloids - colour of petaloid is a noticeably different from petal. Sometimes main petal also has contrasting colour on reverse, but not always

2.Chimeras - usually moderate sized petaloids tucked inside involute petals that cup around them, but some fold back. Distinguishing feature is that blooms can present with different colours on same plant, wolf petals or half/half colouring.

3.Textural - "Ragtag/ruffle/messy" form, some with twisted or ruffled petals, some with split tips. Petaloids present but always same colour as petals and add to chaotic texture rather than being distinct.

4.Decorative form with same colour petaloids - recognisable decorative form with same colour, moderate sized petaloids. Tend to have good bloom depth and so many petals that some present with hard centres early season.

5.Large layered petaloids - distinguished by fully developed petaloids (sometimes extending further than the petal itself) that mimic petal colour. Petaloids are functioning more like petals than the other variations. My “Romance” line.

6.Whisper line - blousy blooms with relatively shallow depth and open sparse petal arrangement. Some petals have a twist to them. Petaloid consistency varies between blooms. I’m unlikely to pursue this line for cut flowers, as it lacks vase life and some other functional faults. Holding on to this line for colour and diversity reasons.

'Lileah Pink Petticoats' surrounded by progeny (from seed & pollen, including second & third generations)

Petticoats, alicorns and chimeras

When Kate Mungoven from Lileah Dahlias sent me 'Lileah Pink Petticoats' several years ago, my initial focus was on the presence of petaloids in those blooms.

However, after the first round of seedlings, my interest shifted to the fascinating genetic colour inheritance that was often passed on (both via seed and through pollen). The curiosity that those tonal variations sparked in me has led me to explore pigment production and colour theory in dahlias more deeply.

'Lileah Pink Petticoats' (previous page, centre) is magenta pink with random vertical orange stripes. Interestingly, seedlings inherited that same pattern in varying saturations of colour. As you can see from the photos opposite, this ranged from the lowest saturation (white/yellow iteration) through pastel (soft pink/apricot) mid (pink/mandarin), the original (magenta/orange), to full deep saturation (plum/burgundy).

In particular, seeing the white with yellow version was a lightbulb moment for me. The independence between an underlying pattern and overlay of pigment saturation suggested to me that there might be different genetic instructions for each of these aspects. It was time to do some deeper research into dahlia pigment production. Whilst I could just continue breeding by putting "like with like" and relying on practical experience and observations, understanding the biology would help me hybridise these lines more intentionally and work toward more precise goals. I've summarised what I discovered about pigment pathways on page 35.

With this particular line, my research also led me to periclinal chimeras. The term sounds complex, but the concept isn't. There are different layers in a petal - the outer surface layer (L1), and two deeper layers (L2 & L3) and in some cases different layers can have different genetic instructions for pigment. This presents visually as a chimera, with random patches of colour presenting in a single layer.

So in this case it's quite likely that one layer (possibly L2) contains transposon "jumping genes" that result in the white with little breaks of yellow. In another layer is different genetic coding instructing anthocyanin pigments to be produced.

In typical dahlia style, we get a large diversity of pigment instructions passed on when seedlings are grown, which is why sometimes the seedling will have an small amount of pigment, other times more. And sometimes that pigment production is silenced completely, giving a clean white "filter" in that outer layer.

'Alicorn' goal: strategies and outcomes

After discovering this unusual colouring, a new goal I pursued was the soft pastel iteration that reminded me of my niece's "Alicorn" soft toy. (apparently an "alicorn" is a unicorn with wings - it seems fantasy creatures have expanded somewhat since the '80s). However, the functional traits of the seedlings that I had with this colouring were very poor. I wanted to capture the colouring but improve the stem strength, bloom attachment and position, and the petal count. I also wanted a neater, tighter bloom form - something along the lines of a ball type.

My hybridisation strategies in the 2024-25 year included:

- Hand cross between L'ancresse x Alicorn (30 seeds)
- Hand cross between Serenade Buttercream x Alicorn (14 seeds)
- Open pollination collecting seed from the Alicorn seedling (pollen parents mostly siblings and other first year seedlings 200+ seeds)
- Returning to the same seed parent OP 'Lileah Pink Petticoats' (200+ seeds)
- Sibling crosses from the first year seedling hub (seed from the 'Lileah Pink Petticoats' first year seedlings block (OP)- 100+ seeds)

Although there were a handful of seedlings worth keeping from all of these blocks for other reasons, from almost 550 seedlings, only two were a clear step forward in this specific goal in terms of both colour and functional traits. One of the keepers was from L'ancresse x Alicorn, and the other from the 'Lileah Pink Petticoats' sibling-cross seedling hub. I hope to further this goal again next season.

Left & below left: original 24-704LPP 'Alicorn' seedling
Below, centre: 25-188LxA - first year seedling L'ancresse x Alicorn
Below, right 25-185LPPHub - first year from LPP sibling cross hub

Dahlia colour pigments

In the cells of petal tissue, raw chemical ingredients are transformed into pigments which result in the colours we see.

The presence, absence or combination of the following pigments contribute to the incredible diversity of colour in dahlias:

- Chalcones: including Butein for bright saturated yellow
- Aurones (derived from chalcones): including Sulfuretin for strong orange-golds
- Anthocyanins: Cyanidin for magenta, purple, burgundy red and Pelargonidin for salmon, orange, bright reds
- Flavones: co-pigments that themselves are sometimes colourless, compete with anthocyanins and so either soften those colours, or other times intensify them, plus can impart slight pale yellow/cream colour

The biosynthetic process that is responsible for the production of pigment is called the flavonoid pathway.

It is directed by the genetic code which provides an instruction manual for which pigments should be manufactured in this assembly line, and the "master switch" of this is an enzyme called chalcone synthase (CHS).

There is a limited amount of precursor (or "ingredients") to create these pigments, so there can be competition for those resources. Sometimes these instructions can be either silenced or regulated, and pigment production can be disrupted, which will result in white petal tissue, or a shift in the typical colouring of that particular cultivar.

'Serenade Tiny Dancer' line for wedding pastels

I've been searching for many years for small ball-adjacent dahlias for my wedding clients in soft neutrals, ballet pink and warm, soft, coffee tones. I hadn't found what I was looking for within current Australian cultivars, so the obvious solution was to hybridise for it. It turns out this was easier said than done.

I had tried collecting a lot of open pollinated 'L'ancresse' seed, and never had a single seedling worth keeping. I also tried hybridising from brighter pink balls like 'Sir Richard' and 'Joan O', but the saturation was too high in the resulting seedlings, so I didn't get any closer with those seed parents.

'Serenade Fairylights', due for release this year (below, left), is gorgeous but more decorative in form and looser petal arrangement. My wedding florists do absolutely love it and I'd like to compliment it with something tighter. 'Serenade Queen of the Faeries', a delicious soft warm pink, is another favourite decorative that is coming through my program (heading into third season).

Below: left - 'Serenade 'Fairylights' and right - 'Serenade Queen of the Faeries'

Above: left - 'Serenade 'Tiny Dancer' and centre & right - 25-186TD seedling

However, this still lacks the tighter spherical form that I am envisioning. 2024 gave me 'Serenade Tiny Dancer' (top left) from open pollinated white pompon seed ('Bowen'). The clear clean pink micro is a sweet little dahlia that gives an abundance of blooms that I love popping into my florist buckets. It was the next generation of seedlings from this as a seed parent though, where the magic really started to happen.

I collected open pollinated seed from 'Serenade Tiny Dancer' (a first year seedling that Autumn), and the results were such a delight to discover this season. Although the pollen parents were all first year seedlings themselves, the overall calibre of next generation seedlings was excellent, and in particular, there was one that I'm very excited about graduating to its second season. (25-186TD centre and right, above). It is everything that I had imagined for my goal and more. It is exceedingly floriferous, and has exceptional health. The plant is self-supporting, and has all cut flower functional traits to the highest level. It even produced an abundance of tubers, pollen and seed, so I'll be able to explore the next generation this coming season and push into this line even further.

I've taken dozens of Autumn cuttings. I would be devastated to lose it so I wanted insurance, plus I'm hoping to grow out a large block to use production and also for assessment next year. I can't think of any seedling I've assessed that has scored so high in its first season. If I had to name a fault, the very first flush had the slightest hard centre. However, that resolved as it opened, and being gold rather than green, it really didn't mar the aesthetic.

Childlike delight - an unexpected giant

It is such fun when something appears that is completely unexpected. One such dahlia was this fiery giant that popped up in a block that I had planted with a very loose goal of breeding for "black" dahlias. I've since learnt a lot more about how one might pursue that goal (intentionally building lines with the highest levels of anthocyanin will give darkest results). I've decided to let go of that goal for now (one can only do so many things at once!).

That year though, I simply collected seed from a very dark burgundy miniature decorative out of curiosity. Given the proximity of my 'Hamari Gold' plants and the strong parallel in form, it seems likely that 'Hamari Gold' was the pollen donor.

What surprised me was that, contrary to what I'd heard, giant size can come from quite a small parent. It was possibly a testament to the depth of line breeding behind 'Hamari Gold' and many copies of genes for large size that was passed on.

The working name for this bloom is 'Serenade Firedance'

Above: all second generation seedlings from 'Serenade Firedance', below

At the end of that first season, I introduced pollen from 'Serenade Moondance', and also left those same blooms to open pollinate. This means I don't have any certainty to the pollen parents, but I ended up with a good portion of seed.

The next generation was consistently large. This was again impressive considering much of the pollen came from smaller blooms. There was also a broad diversity in colourings. (See above for some of the resulting seedlings).

If breeding giants was a specific goal of mine, I'd certainly be crossing these with other giants, even perhaps backcrossing to 'Hamari Gold' to further that line. However, this has become just a fun little indulgence. Giants don't tend to have a great vase life, so they're more of a novelty than a practical cut flower.

Three seedlings: (L-R) 24-725Moon, S. Moondance, 25-018(725)

Dancing in the Moonlight..

Several years ago, a seedling was produced from open pollinated 'Coorabell Jessie' seed that was quite different to the others in its block. I suspect the pollen parent may have been a large white waterlily-adjacent bloom 'Joy Burkitt'.

That seedling, 'Serenade Moondance', (above, centre) is the perfect aesthetic for weddings, with clean white ruffled petals. Functionally it has strengths of an exceptional cut flower - better than either of its parents. The stems are some of the strongest, longest and finest (not chunky or hollow) on the farm. The blooms have a very secure attachment point to the stem, and they present 45° to upright position atop that strong stem. Especially considering the size of the blooms they do have a good vase life, and the petals do not bruise easily. Best of all when growing in our hot, humid, subtropical climate, these white dahlias have never discoloured or sunburnt in the five years I have grown it.

I've found that these traits tend to be passed on to the seedlings very readily. Pictured above are three generations. 'Serenade Moondance' centre, seedling to the left and that seedling's seedling to the right.

This line of seedlings is hands down the best I have assessed in any of my breeding patches. It has been so gratifying to see the results of building up through the generations of my own lines, and the result of being disciplined with removing anything that does not align with my functional goals each year.

Whilst of course there were still some seedlings that were removed, I was very discerning with my assessment, and feel confident with the exceptional calibre of the graduates from this year's breeding.

In addition, I'm seeing some of the same strong traits being inherited via pollen. This year I continued working these lines, and the unusually fine April weather gave me the opportunity to use exclusion bags and hand pollinate. I made some back crosses to the seed parent and brought in pollen from other functionally excellent blooms with colouring or traits I wanted to explore within this line. The seed from this next generation is some that I'm most excited about for next year.

This page: All seedlings from Serenade Moondance OP seed

Above: Seedlings inheriting 'Serenade Moondance' traits through pollen

Below: third generation 24-725Moon with its own open pollinated seedlings all around it

Surprise from 'Oddity'

I grew out open pollinated seed from an unusual dahlia 'Oddity' bred by Jill Hogan from Red Rabbit Blooms. It has an anemone-like centre, with multiple back layers of petals presenting with tiny petaloids.

It was a random roll of the dahlia genetics dice, and I was of two minds about germinating this small handful of 'Oddity' seed as I approached my self-imposed 2000 plant limit. However, as usual, curiosity got the better of me, and I'm so pleased it did. Unfortunately, being a bit of an afterthought, this block was planted in the least desirable bed in the seedling patch. Clearly if they made it through assessment in those challenging conditions, I expect them to perform even better next year.

Above: 'Oddity' (both images)

My first year seedling (left) that came from this seed with contrasting colour petaloids and also darker reverse petals was functionally passable, but with aesthetic traits too tempting to not keep for breeding. I'll be working to further improve the stem strenth, attachment and position of this line in the coming seasons. The pollen parent was very likely another petaloid bloom sent to me by the same breeder which also has the darker petal reverse which was growing next to 'Oddity' the previous season

Far outside the standards of any recognised form was this bizarre seedling. The disc centre was very large and dominated the bloom. The ray florets around the edge were fused tubular petals, echoing the inner section of an anemone bloom, but they curved dramatically inwards. I'm a little hesitant to pursue this as a new direction because of the open centre, but I couldn't resist tagging this to keep, and also collecting a small amount of seed... just to see what might happen.

My favourite by far is this seedling with dramatic contrast in pigment between front and back of the petals. It makes for a very moody aesthetic. It is also has some petaloids. The working name for now is 'Serenade Baroque'.

Although there are some cultivars that present like this overseas, as far as I know, we haven't had access to anything with quite this stark contrast in Australia to date. Like the petaloids, this seedling would be selected against in exhibition breeding. With its "in-betweener" form, it sits more comfortably in the realm of cut flowers or garden cultivars.

Functionally, this seedling wouldn't pass if being assessed for stem alone, as it isn't as strong as I require from my cut flower cultivars, but this may be a result of the poorer growing conditions. I look forward to seeing it in its second year.

I did some hand pollination with this seedling, trying a few different strategies.

- 'Breannon' pollen for reverse contrast and good functional cut flower traits.
- Two-way hand cross with my best cut flower seedlings that have similar form
- Back cross to the most likely pollen parent, given its proximity and traits including contrasting reverse petal colour and petaloids.

I also left some blooms to open pollination, as I do sometimes still like to roll the dice and see what else might arise, especially in the more experimental lines. This will be very interesting seed to grow out next season - can't wait to see if some of the next generation share that dramatic pigment

A cut above - productive farm lines

First year seedlings are fresh, new and completely unique. We can be blinded to their faults, or make excuses, hoping those things might improve. It is in subsequent year evaluations where things get really interesting. The novelty factor has softened, and we observe them in larger numbers, comparing them alongside tried and tested cultivars. By second season I have very high expectations, and with more plants to observe, faults are harder to ignore.

As already mentioned, my minimum requirements for cut flower lines to graduate:

- STEM: strong, long, straight and fine.
- ATTACHMENT: secure, absolutely no movement when I give it a wiggle.
- POSITION: 45° upward or higher, no clock-faced forward blooms.
- CENTRE & PETAL COUNT: dense, fully petalled and ideally not hard, malformed centre, plus good petal retention (holding petals when fully open).

Additionally I like to see

- HEALTHY & RESILIENT: no dahlia divas on my farm, and they need to perform under the stress that our climate invariably throws at them.
- FLORIFEROUS: Productivity is essential in cut flower farming. With our long season, early, mid and late flowering types are all useful, but if they are workhorses for the 4-5month flowering season here, all the better.
- CONSISTENCY: Bloom colour or even size variations are fine, but I need the majority of the blooms to be A-grade, so the quality needs to be there consistently. Early season wonky form, uneven hard centres or other persistent aesthetic faults are not tolerated for long.
- ROBUST PETALS & LONG VASE LIFE: Tight form and petals that won't bruise or discolour easily, and they need to last in a vase.
- SELF-SUPPORTING & GOOD HABIT: With minimal plant supports in place, I need plants that won't topple easily and have low centre of gravity that don't need assistance staying upright. Long stems, but originating deep in the plant.
- AESTHETICS: I'm looking for colours and forms that suit my client base.

On the following pages are the best from the cut flower lines being assessed in the main cutting patch. All score highly in all of these areas, and they provide a range of colours that suit my retail florists and wedding stylists.

24-531DS - 'Serenade Katie' (left)
Dreamy neutral colouring, perfect size for featuring in an arrangement, and it sits so nicely on the stem. The seed parent was one of my favourite named cultivars, 'Devon Salmon' (below).

24-744BS - 'Serenade Queen of the Faeries' (left)
Exquisite colour, I adore this seedling so much that I named after my Mum. The seed parent was 'Boundary Sophie', (below).

24-600vol (above)
Warm soft pastel that works beautifully with other low saturation colours, and has a long harvest window, giving flexibility of 3-4days at optimal stage for cutting for good vase life but also looking full and open.
The Seed parent is unknown.

24-726CKD (above) A soft pink which is very popular with both wedding stylists and my retail florists.
The seed parent was 'CK Dream' (above right), a densely petalled decorative that has passed on all its strong functional traits to this seedling.

24-863(062) (above)
This seedling is a second generation Serenade and excellent cut. The seed parent, (right) sadly has some issues with sunburn on white petals, but has produced a great line of seedlings.

24-482LPP (above)
This seedling is from the line discussed earlier with the interesting chimera inherited from 'Lileah Pink Petticoats' (left). This third year seedling has exceptional functional traits and is the best of the pink/orange seedlings for cutting. Sports to an all-orange version at times so will continue to observe and watch for consistency.

24-585Isa - 'Serenade Golden Slumbers' (above) is a perfect cut flower in a lovely saturated peach with gold tips. Needs to slightly improve productivity to secure its place in the cutting patch. The seed parent was 'Isabel' (above right).

24-806(247) (far right)
Second generation Serenade seedling which has performed exceptionally. I am dropping the seed parent (below) due to forward position and inconsistency, but the new seedling has inherited a similar colouring and is functionally so much better.

The year before last I grew out over 200 seedlings from 'Blyton Softer Gleam' open pollinated seed. With so many functionally excellent seedlings, I kept over 20% of these to look at for a second season. The three standouts you can see on this page. I am still keeping several others to assess for a third year.
'Blyton Softer Gleam' directly below.

24-596BSG - 'S. Cloud Dancer' (above)
I always need more whites for our wedding clients. This has a touch of lilac and a lovely sense of movement like whispy clouds to its upturned petal tips. It seems that the pollen parent may have influenced form and colour more than in other seedlings.

24-770BSG (above)
One of the few Serenade seedlings that potentially meet show standards. It conforms to the ball form and at the same time is a successful cut flower. It has a warm vibrancy that works in brighter palettes without being overbearing.

24-712BSG (above)
Bronze gold with slight darker reverse to the involute petals. This seedling does cut well but the plant isn't quite as self-supporting as I would typically want. Otherwise excellent.

Other 2026 seedling notes...

Brown and other muted tones of pink, peach, rust and bronze are sometimes requested by my floral stylists. I'm not actively breeding for this colour, but if such a dahlia appears it deserves a second look. This one also assessed well on all functional criteria. This seedling is second generation Serenade, the seed parent a dusty pink (inset) itself grown from from 'Devon Salmon' seed. As well as the colour, I also love the petal density and arrangement of this new bloom, and the overall proportions and shape. The only thing I'll be watching for next season is productivity.

24-761(129) and next generation

I discovered an interesting soft lemon seedling two years ago which featured a significantly larger back row of petals. The form suggested possible gene inheritance from an anemone type. I was intrigued by it, and thought I might explore the line a little further by collecting open pollinated seed.

There were 2 new seedlings this season that inherited a similar form (both inset). The soft pink one was very close to the original, just in a different colour, and I've kept it for future observation. The pinky-peach was more pronounced separation between the back petals and central ones and lacked functional traits to make it useful for cut flowers, and I've let that go.

25-080(604) I don't generally gravitate to variegated dahlias with dark pigment breaks. However, this soft lemon-beige with subtle burgundy flecks gives delicate interest to a pastel palette. This seedling self-seeded underneath the only speckle seedling that had come through the previous season. (inset)

Another self-seeded volunteer in the first year seedling patch was this lovely apricot. Exceptionally productive and a lovely even soft apricot that held its centre late into the season, I'm predicting this will become a real workhorse and a staple in many of my florist buckets in the coming years.

Autumn

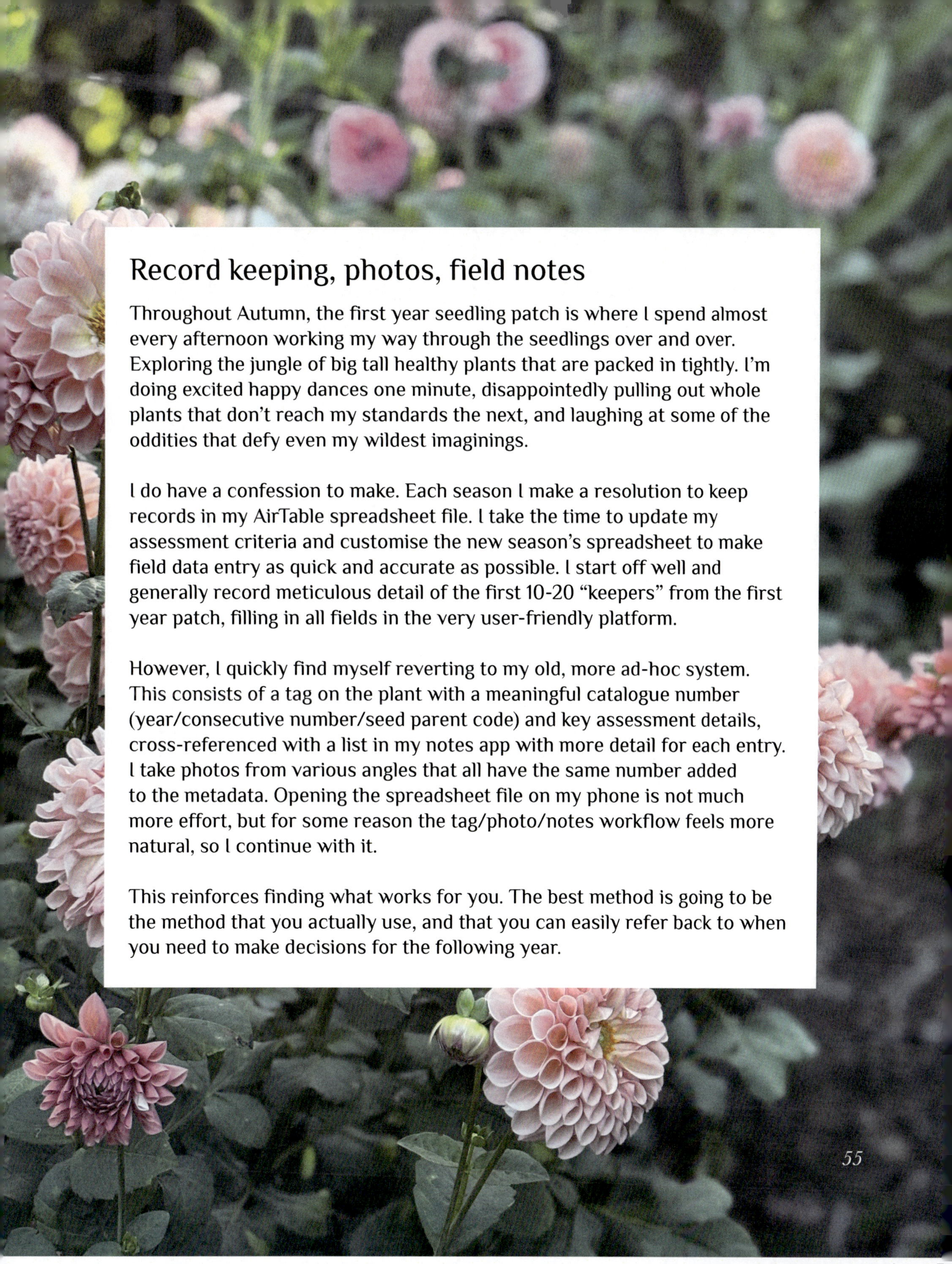

Record keeping, photos, field notes

Throughout Autumn, the first year seedling patch is where I spend almost every afternoon working my way through the seedlings over and over. Exploring the jungle of big tall healthy plants that are packed in tightly. I'm doing excited happy dances one minute, disappointedly pulling out whole plants that don't reach my standards the next, and laughing at some of the oddities that defy even my wildest imaginings.

I do have a confession to make. Each season I make a resolution to keep records in my AirTable spreadsheet file. I take the time to update my assessment criteria and customise the new season's spreadsheet to make field data entry as quick and accurate as possible. I start off well and generally record meticulous detail of the first 10-20 "keepers" from the first year patch, filling in all fields in the very user-friendly platform.

However, I quickly find myself reverting to my old, more ad-hoc system. This consists of a tag on the plant with a meaningful catalogue number (year/consecutive number/seed parent code) and key assessment details, cross-referenced with a list in my notes app with more detail for each entry. I take photos from various angles that all have the same number added to the metadata. Opening the spreadsheet file on my phone is not much more effort, but for some reason the tag/photo/notes workflow feels more natural, so I continue with it.

This reinforces finding what works for you. The best method is going to be the method that you actually use, and that you can easily refer back to when you need to make decisions for the following year.

Consistency

Sometimes I'm confident to cull after the first bloom. If there are simply too many faults, such as a weak stem, low petal count, or a colour I'm not interested in, I remove it immediately. However, most times I give the plant at least 2-3 flushes to improve. In my experience, most often it won't.

On rare occasions though an ugly first bloom does improve substantially throughout its first season. The series of photos above shows different blooms from the same plant, the first on 14th February through to the last on 30th March. What the photos can't capture is the significant improvement in stem strength and bloom attachment over that same period.

Hybridisation

By early to mid-April we had finished our cut flower orders for the season, freeing us up to switch to a full-time hybridisation focus. A delightfully unseasonal dry start to Autumn allowed the best opportunity we have ever had for hand pollination. With organza bags and paintbrushes in hand, twice a day I collected pollen, moving between selected seed and pollen parents, pretending to be a bee! It was time-consuming, but I wasn't going to squander such a rare opportunity to have real control over which cultivars were being crossed. My success was mixed. Some cultivars resulted in more seed than I expected, whereas others didn't result in any.

To hedge my bets I also implemented my usual strategies. These included rubbing ripe blooms with visible pollen over desired seed parents, cutting back undesirables to let the bees do the work, and of course all the winter and spring planning of hybridisation hubs and close plantings, which I had already set up for selective open pollination.

Above: Large Organza bags are such a simple way of excluding pollinators from blooms.
Opposite: When the dahlias completely open in Autumn they will present the disc centre. Over several days you will see individual florets ripen (from the outside of the disc inwards). They first present anthers with the pollen, which you can collect, and then you can apply that pollen to the desired seed parent when they are showing stigmas (which look like tiny yellow 'Y" shapes).

Hand Pollination in photos

Seed ripening

Of course then our Autumn rain came (and came, and came), but we didn't panic and race about harvesting immature pods. On the occasional dry and sunny day we would harvest only the fully mature pods, as viability and germination rates improve substantially when seed is allowed to ripen fully on the plant. With no frosts right through May and June, we're fortunate to be able to just let them ripen naturally which ensures optimal viability.

This page and inset: dry pods ready for harvest
Opposite: Winnowed seed ready for hand grading

Above left: harvested seed sorted by seed parent.
Above right: handmade winnower - free DIY plans at www.realseeds.co.uk

Seed drying and grading

May & June sees us busy processing, packing and dispatching our Serenade seed. Once harvested, we allow it to thoroughly dry on racks, then run through a DIY seed winnower to get most of the chaff off, and then sort, clean and grade by hand.

Whilst some hybridisers very understandably keep all their seed stock for their own growing, I'm happy for others to share in the joy of discovery that comes from growing from seed. There's simply so much genetic variation to go around. However, I do encourage growers to engage in all the other steps: creating their own goals, planning hybridisation plantings and crosses, collecting seed and selecting from their own lines. However, if this isn't possible, growing seed from a devoted dahlia breeder who takes such care with all of these steps can be the next best thing.

62

Autumn cuttings

Cloning our best seedlings is the way we guarantee their survival in future years. Tuber division is the most common way of doing this. It is possible, however, that a seedling's single tuber clump might succumb to rot or poor storage then it will be lost forever. So if your climate is suited to it, or you have an indoor winter growing space, it can be protected by taking cuttings during late Autumn and over-wintering those.

I will often take cuttings from my very favourite first year seedlings, as well as second year seedlings that I have loved and want to grow in much higher numbers the following year. This does mean I need to keep these alive through Winter. I begin this process in a warm insulated grow room with supplementary lighting, and once established, I bring the plants out into our beautiful winter sunshine. I then often use these as mother plants to take more cuttings in late winter, to be planted out when the soil warms in Spring.

Alternatively, if they have sufficient growth I'll sometimes acclimatise them to winter conditions, let them grow a while longer through the short days, then dry them off for a brief dormancy before planting their little pot tuber clumps whole into the soil alongside the rest of my tubers.

Winter

Philosophical reflections and evolution

As we enter Winter, writing this book has allowed me to reflect on the season that was. I'm gaining clarity that there are two distinct arms to my breeding journey.

First of all, there is the more straightforward process of hybridising and evaluating seedlings with the goal to create exceptional cut flowers for production on the farm. This feels more akin to engineering, even with dahlia's genetic diversity and unpredictability, and is fairly well aligned with hybridising for exhibition blooms, but just using different assessment criteria. I have a fairly clear idea of what I would like to achieve, I learn as much as I can about the underlying science, and then I take steps to hybridise cultivars to hopefully result in something close. Each year I refine that until I am happy with the result.

At the same time, but in separate genetic lines, I am still very much rolling the dice, and discovering what is possible with this fascinating genus. Pure curiosity. I imagine what it must have been like to be a botanist in the early 19th century, growing seed from some of the very first hybrids and discovering new variations with every season. I feel that this process shouldn't be rushed, and my mind needs to stay inquisitive and open to possibilities.

I don't want this second goal to be entirely random and unfocussed, otherwise it is an aimless brainstorm without any forward movement. So I'm making sure to observe closely. I have been recording the detailed nuance between traits in these unique forms, so that I can begin carefully crafting new lines with distinct identities, without rushing them to market simply because they are different to what is currently available.

Ultimately these two goals may merge, as the functional traits of my more experimental lines gradually meet the standards of my cut flower requirements. For now they can happily co-exist, and I can work as both engineer and artist in my dahlia breeding.

Another evolution in my breeding philosophy is that I'm more comfortable with pivoting my goals and focus than I was initially. I don't need to rigidly hold tight to a specific outcome. I also don't feel like I've failed if I don't achieve a specific dahlia that I may have dreamed up. I might begin with a certain idea in mind, but it remains theoretical until I start to see it take form. Season by season, as it becomes partially realised, the goal itself often shifts slightly. And that's fine.

I'm also happy to be taken on fascinating little side trips when something unusual arises and suggests a slight detour to my initial plan. With limited time, energy and growing space, this sometimes means retiring old ideas or directions that aren't bearing fruit. This makes room for a new dream.

Farm News

Serenade Farm does not exist in a vacuum. Beyond the farm and the turning of the seasons, we have an influence on the wider world around us, and it on us.

A significant part of the personal meaning and purpose I derive from dahlia growing comes from the community around me - both online and in person. I love the teaching side of what I do, and sharing knowledge and experiences with others. So this seasonal journal would not be complete without some of the key events that happened since the publication of "Cultivating Curiosity" last Winter.

Dahlia Society of Australia conference

My second national conference was held in June 2025, and I loved every moment. There is truly nothing like having an entire weekend devoted to the dahlia. There were endless conversations with people from all walks of life who are as obsessed with dahlias as I am, plus carefully curated presentations and sessions that both support and validate my own experience, and at the same time challenge and expand my knowledge base. It was such an honour to join other Australian hybridisers on the panel session on dahlia breeding and share experiences with our varied goals, strategies and evaluation criteria. Hands down the highlight for me personally, was hearing Dr Keith Hammett present his keynote, and to have the opportunity to have several conversations with him about my own breeding goals, and process. Having devoted decades of his life to plant breeding, he was a genuine wealth of knowledge, and the joy he still so clearly derives from it was truly inspiring.

International podcasts

After publishing "Cultivating Curiosity", I was invited to speak on two international podcasts in the last year. Both were really affirming experiences, and gave me the opportunity to reflect and formulate answers to some genuinely interesting questions about my specific practices.

I spoke with Joann & Allison from "The Dig on Dahlias" from the US. Being a podcast devoted exclusively to dahlias meant that the audience was already very well informed, and it was great to really get into the weeds rather than only skimming the surface. We had such fun chatting all things dahlias and breeding, and comparing notes on just how different our climates are — subtropical Queensland versus the Pacific Northwest in the opposite hemisphere.

Ali from "The Dahlia Dialogues" broadcasts on Instagram and is based in Kent in the UK. Her audience also already has a great foundational knowledge of dahlias, and I absolutely loved the conversation, which explored the nuance of how I do things here and why. I was also delighted to discover that Ali has studied neurophysiology. This reassured me that at least one reader truly appreciated the little section I couldn't resist including in "Cultivating Curiosity" on dopamine and how dahlia breeding's tendency for reward prediction error keeps us so happily hooked.

Serenade release plant sales

2025 saw the first official Serenade cultivar release!
In collaboration with trusted friend, Mardi Rhoden, from Quite Contrary Flower Farm, we released five Serenade Dahlias as garden-ready plants.
Mardi trials many of my seedlings in her much colder climate each year to give a more rounded assessment. In late winter and Spring each of us raised hundreds of cutting-grown plants and then during November we held sales in Tamborine Mountain (where Serenade is based) and Labertouche (Quite Contrary's town in Victoria). The five new cultivars are shown on the opposite page.

Dahlia devotees travel ridiculous distances to be there, and it was such a buzz to meet so many lovely growers on the day. We had amazing friends by our side helping out with order pickups and extra sales, and I also ran a free propagation workshop to demonstrate growing from cuttings, which was really well received.
Mardi is a dream to collaborate with, and our partnership continues this year. She has been growing out all of last year's releases to offer as tubers this Spring, making them available online, alongside a few special new releases.

Queensland Dahlia Society state show

The QDS are such a lovely group of people. I've connected with them several times over the last few years and they are welcoming, supportive and inclusive. For the first time this year they opened up new categories specifically for cut flower farmers.

The cut flower categories had plenty of entries, and it was tremendous to see the quality of blooms and cut flower dahlias so well represented. It will be great to work further with the society to refine judging criteria to reflect traits that matter most to the cut flower industry as well.

Gardening Australia TV filming

More exciting news from this season was an invitation to appear on ABC's beloved national program, "Gardening Australia". It was such a lovely surprise to be contacted by the producer, and it all happened in a flurry less than a week later — with the production crew and the wonderfully knowledgeable Jerry Colby-Williams visiting us on the farm for a long but wonderful day of filming. We are yet to hear what date the episode will air, but hopefully in the coming months.

Australia Post dahlia stamps

It was such a joy to see some of the dahlias bred and grown in Australia celebrated by Australia Post with the release of a series of dahlia stamps in May. The designers consulted with the very knowledgeable Deb Collett who ensured all the dahlias represented in the stamps and associated artwork had originated in Australia. We were fortunate to be asked for the use of one of the photos of our Serenade cultivar, 'Serenade Whisper' to appear in the marketing artwork, including the front page of the mini sheet and first day release packs.

What's next?

The Australian national dahlia conference is happening again this month, this time in Coffs Harbour. I'll be presenting a session on "Breeding Better Dahlias" and diving into how to rigorously assess for quality cultivars, even when working toward goals outside the traditional forms and standards.

Winter will be the release of this book alongside some delicious seed packs and bundles, including some special lines that we haven't offered before.

And then in no time at all, we'll be into the next season cutting hundreds of stems from dawn each day for our lovely florist's orders, and discovering the next generation of Serenade seedling treasures. I will continue with all the lines that I've discussed in the "Summer" section of this book, pursuing the new discoveries that appeared this year, and looking out for more surprises.

And so the seasons will turn, and in each and every moment so much to be grateful for - the anticipation, discovery, joy, reflection...

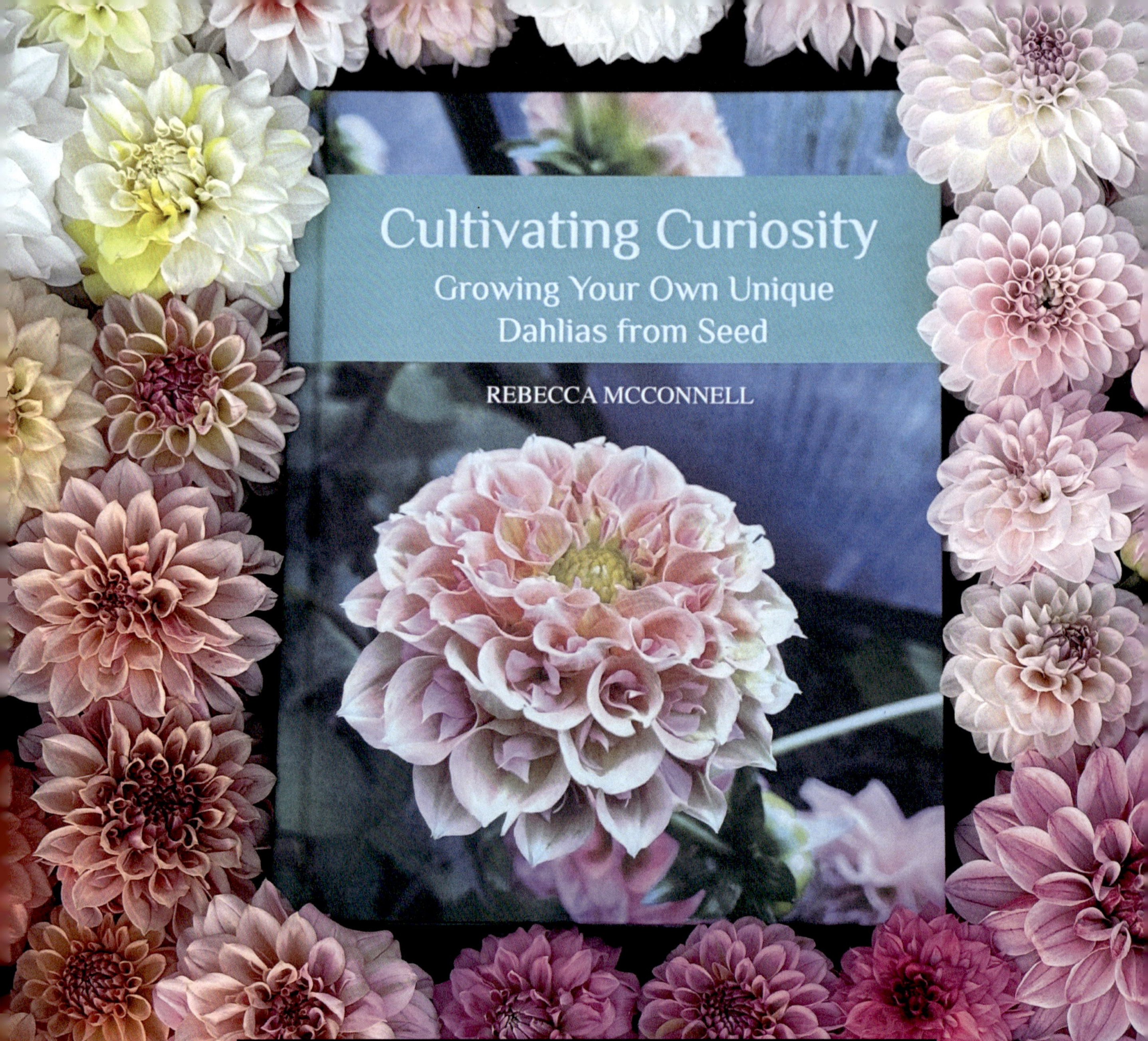

My first dahlia breeding book, which delves into general principles for growing, assessing and hybridising strategies is called “Cultivating Curiosity”, self-published 2025

www.serenadefarm.com.au
Our dahlia cut flower farm & breeding program is based on Tamborine Mountain, Australia

“Beyond Curiosity” was first published June 2026

Made in United States
Orlando, FL
17 September 2026

87862214R00045